**Modular Science**

# HEAT & INSULATION

## Ingleby Kernaghan

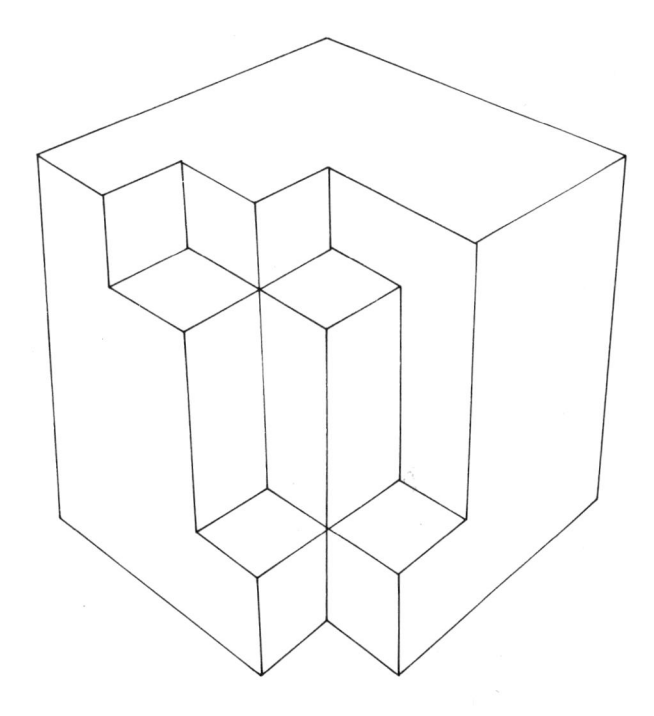

Blackie

Glasgow and London

**Acknowledgements**

The author and publishers wish to thank the following for permission to reproduce copyright material:

P. W. Allonby (page 5); Robert Hamil (page 6); Science Photo Library (page 11, left); British Broadcasting Corporation (page 11, right); G. Maunsell and Partners, Consulting Engineers (page 12, left and top right); B.P. Oil Limited (page 12, bottom right); Honeywell Control Systems Limited (page 14); Bryan and Cherry Alexander Photography (page 18, far left); Barnaby's Picture Library – Hubertus Kanus (page 18, middle)/Zita Blackburn (page 18, right); The Royal Society for the Protection of Birds and G. St. J. Hollis (page 19); Thermos Limited (page 21); Tube Investments Ltd. and British Gas Corporation (page 27, top); Valor Heating Ltd. (page 27, bottom left); Ouzledale Ltd. and the Solid Fuel Advisory Service (page 27, bottom right); the Welding Institute (page 28, left); The Natural Energy Association, London (page 28, right); Central Office of Information (page 30); Potterton International Ltd. (page 31).

**Modular Science**

This series was planned and developed with the assistance of an Editorial Panel. Its members are:

Bob Fairbrother, Lecturer in Science Education, Chelsea College
Edgar Jenkins, Senior Lecturer in Science Education, University of Leeds
Peter Scott, Headmaster, City of Leeds School

ISBN 0 216 90586 9

First published 1981

Copyright © Ingleby Kernaghan 1981

All rights reserved. No part of this publication
may be reproduced, stored in a retrieval system
or transmitted in any form or by any means,
electronic, mechanical, photocopying or otherwise
without prior permission from the publisher.

Published by Blackie & Son Ltd
Bishopbriggs, Glasgow G64 2NZ
Furnival House, 14–18 High Holborn, London WC1V 6BX

*Phototypeset by Filmtype Services Limited, Scarborough*

*Printed in Great Britain by Bell & Bain Ltd, Glasgow*

# CONTENTS

*The work in the first part of this book is called the core. Your teacher will probably want you to work all the way through it. It contains the following chapters:*

**CHAPTER 1**
Keeping warm and keeping cool

**CHAPTER 2**
How does heat travel?

**CHAPTER 3**
Heat and expansion

**CHAPTER 4**
Heat and insulation in the house

**CHAPTER 5**
Clothes

**CHAPTER 6**
Keeping food hot or cold

**CHAPTER 7**
Temperature and heat energy

*When you have completed the work of the core, you should know quite a lot about the properties, effects and uses of heat. You should appreciate the importance of heat and the value of saving it.*

*Your teacher will then want you to do one or more of the following options:*

**OPTION A**
How do we get heat?

**OPTION B**
Fuels for home heating

**OPTION C**
Heat saving in your home

**OPTION D**
Home and hot water heating

*When you have done this you will know even more about how to produce heat and how to 'SAVE IT'.*

# CHAPTER 1
# Keeping warm and keeping cool

## Keeping warm

In winter when you are cold, what do you do to keep warm?

When you sit in front of a fire, the heat is coming to you from *outside* your body. Other ways of keeping warm use the heat from *inside* your body. The food you eat is the fuel your body uses to keep you warm. About three-quarters of your food is used in keeping you warm. The rest is used in doing work.

The heat is carried to your head, arms and legs by your blood, which circulates around your body. The temperature of your body varies a little from time to time, but when you are healthy it is about 37°C (98°F).

A baby, because it is so small, cannot make enough of its own heat, so it must be kept warm all the time by wrapping it up. Some old people also can have difficulty in keeping warm, so it is important that they have warm homes and wear warm clothing. When your body temperature drops, certain things start to happen. This table shows what happens as we get colder and colder.

| Temperature of body | What happens |
| --- | --- |
| 37°C | We are healthy and active |
| 32°C | We feel cold and shiver |
| 30°C | Our thinking is muddled |
| 29°C | Our shivering stops |
| 24°C | We feel miserable and depressed |
| 21°C | Our blood stops circulating |

Shivering is an automatic movement of your muscles as you try to keep warm. If you are weak from lack of food, or if you are very young or very old, you may not have the strength to shiver. The lowest living body temperature ever recorded was 18°C for an American woman who was lost in a snowdrift overnight, and found unconscious next morning.

Every winter many old people die from HYPOTHERMIA which is what doctors call the ill effects you feel when your body gets too cold. Some doctors think that over 50 000 old people die like this every year.

If an old person is found unconscious with hypothermia, the doctor will slowly warm him or her by keeping them well covered in a warm room at 21°C to 27°C.

**ACTIVITY 1.1**
1   Look at the picture of an outdoor skating scene. How many different ways are there of keeping warm?
2   Look at the table and say what you would expect to happen to your body if its temperature dropped to 20°C?
3   The temperature of your room will be about 20°C. Why has your blood not stopped circulating and why do you not feel cold?
4   Why do many old people die from hypothermia every year, when only a few young people do?

## Exposure

We often hear of people who go climbing or walking in mountains dying of EXPOSURE. They can die of this even if they wear the proper clothes.

A mountain rescue team bringing an exposure victim down the mountainside

You are most likely to suffer from exposure in wet, cold and windy conditions, and when you get tired. If your clothes get wet, then the wind blowing over them evaporates the water and you cool down a lot faster (see Activity 1.2). You then suffer the same ill-effects as hypothermia. *To avoid exposure:*
1  Keep warm. Wear clothes made of insulating materials, e.g. woollen shirts and pullovers.
2  Keep out the wind and rain. Wear windproof and waterproof outerclothes.
3  Do not tire yourself out (why?). When you feel tired in this sort of weather, do not go on – stop somewhere sheltered and rest. Eat energy-giving foods such as chocolate.

In Chapter 5 you will see how the clothes of people in cold countries are designed to prevent heat being lost.

## Keeping cool

**ACTIVITY 1.2**
When you get too hot you sweat (perspire). Why do you do this? This experiment should help you to understand why you sweat.
Wet the back of your hand and blow on it. Blow over the other hand, which should be dry. Ask your teacher to put a few drops of propanone (acetone) on the back of your dry hand.
1   What difference do you feel between blowing over a wet hand and over a dry hand?
2   What happened to your hand when propanone was put on it?
3   What happened to the propanone?
4   Explain why sweat helps to keep you cool.
5   What have these experiments to do with exposure?

### Evaporation

The liquid on your hand made it go cold because it was EVAPORATING. Liquids will usually turn to vapour without being boiled. When this happens, heat is taken from the surroundings which are therefore cooled. This is why your hand went cold.

### ACTIVITY 1.3  The thermometer tree

Your teacher may set up this experiment for the whole class, or you may be able to do it yourself.

The thermometers are held loosely so that they can be turned to read the temperature easily. The thermometer bulbs are covered with cotton wool held on with an elastic band. One bulb is soaked in ether, one in water and the last one is left dry.

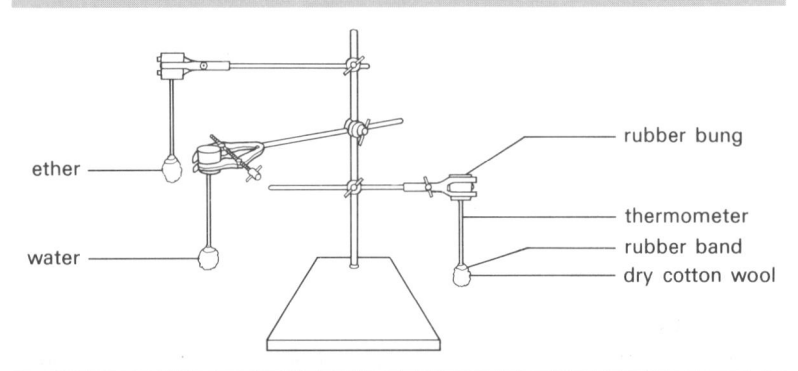

Leave the thermometers set up for 15 minutes, noting the temperature of each thermometer every minute for the first 5 minutes.

Copy this table into your book, and then fill in your results.

| Time (mins) | Temperature |  |  |
|---|---|---|---|
|  | Dry | Water | Ether |
| 0 |  |  |  |
| 1 |  |  |  |
| 2 |  |  |  |
| 3 |  |  |  |
| 4 |  |  |  |
| 5 |  |  |  |
| : |  |  |  |
| 15 |  |  |  |

1  Explain what happened to the temperature of each thermometer.
2 a  What happened to the cotton wool soaked in ether?
  b  Why did this happen?

On a hot day, which is the better drink to cool you down – a cold drink or a hot cup of tea? The answer is – a hot cup of tea! When you drink a cold drink it will cool you a little, but you soon warm up again. The hot drink makes you sweat more and this extra sweat then cools you down by evaporation.

On a hot day people often wear short-sleeved shirts or shorts to keep cool. What they are doing is increasing the area of their skin which is exposed to the air, so that evaporation can take place to cool them down.

Increasing the area of an object will cool it down quicker. Hot soup will cool quicker in a bowl than in a mug. Motorcycle engines have fins to increase the amount of metal which can be cooled by the air.

### ACTIVITY 1.4
Write down some more examples of things that have a big surface area to help cooling down.

# CHAPTER 2

# How does heat travel?

Heat can travel in three ways, CONDUCTION, CONVECTION and RADIATION. The experiments will show you what these different ways are.

## Conduction

### ACTIVITY 2.1  Heat being conducted

Set up the apparatus shown in the diagram. You will also need a stop watch or clock. Stick each ball bearing on with a small blob of Blu-tack. Use as little as possible. The ball bearings should be very close to each other. Start heating the end of the rod with a strong flame, and at the same time start the stop clock. Write down the time when each ball bearing falls.

Be very careful with the rod after it has been heated because it will be very hot. Do exactly as your teacher tells you. Repeat the experiment using rods of different metals.

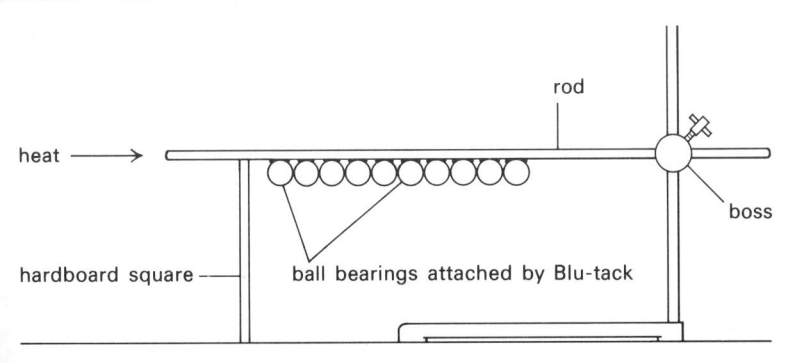

Copy the table and fill in your results.

| Ball bearing | Time | | |
|---|---|---|---|
| | Brass | Copper | Steel |
| 1 | | | |
| 2 | | | |
| 3 | | | |
| 4 | | | |
| 5 | | | |
| 6 | | | |
| 7 | | | |
| 8 | | | |
| 9 | | | |
| 10 | | | |

1  What happened to the ball bearings as the rod was heated?
2  Why did they do this?
3  What is the heat doing in the rod?
4  What were the differences between the rods?
5  Why was the hardboard square put between the Bunsen burner and the ball bearings?

A material that lets heat move easily through it is called a CONDUCTOR. A material which does not let heat move easily through it is called a bad conductor or an INSULATOR.

### ACTIVITY 2.2  Insulators and conductors

Take the temperature of an insulator (e.g. an expanded polystyrene ceiling tile) and of a conductor (e.g. a piece of metal). Hold the insulator against one cheek, and the conductor against the other.
1  What was the temperature of the insulator?
2  What was the temperature of the conductor?
3  Was there a big difference between their temperatures?
4  What was the difference in how they felt on your face?

*Diagram: apparatus showing heat applied to a rod with ball bearings attached by Blu-tack, hardboard square beneath, held by boss.*

5  If they were at the same temperature, explain why the conductor felt colder than the insulator.

Most metals are good conductors of heat, but let us see if a liquid such as water is.

### ACTIVITY 2.3  Is water a good conductor of heat?
Put a small piece of ice about 1 cm$^3$ in a test tube, and gently place a ball bearing on top of it. (If you are using broken ice, fill the bottom 2 cm of the test tube with ice, then put the ball bearing in.) Half fill the test tube with water. Hold the bottom of the test tube in your hand and gently heat the water at the top of the test tube.
CAUTION: Wear safety spectacles. Hold the test tube gently and do not point it at anyone.

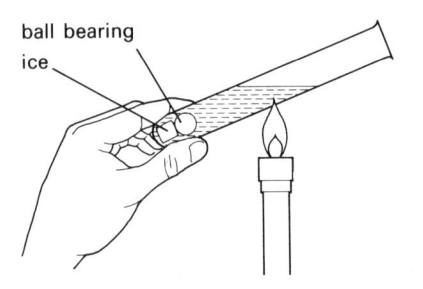

1  Why was the ball bearing put in?
2  What happened to the water at the top?
3  What happened to the ice?
4  Did your hand feel hot?
5  Do you think this experiment shows that water is a good conductor? Give a reason for your answer.

Most liquids are bad conductors of heat, but what about gases? This next experiment shows how air conducts compared with water. The result has important applications in the design of clothes and of houses, as you will see later.

### ACTIVITY 2.4  Is air a good conductor of heat?
Make sure that your boiling tube is dry inside. Insert a thermometer and bung. Write down the temperature. Put the boiling tube into a beaker of hot water and start the stop clock. Stop the clock when the temperature has gone up by 20 degrees.
Now repeat the experiment, but this time start by filling the boiling tube with water which is at room temperature.

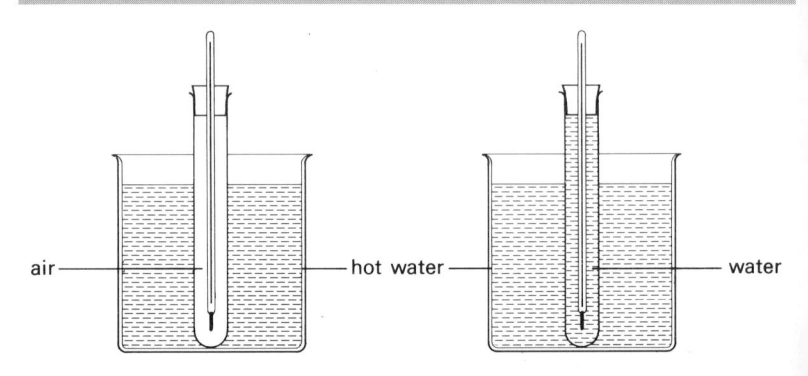

1  How long did it take the air-filled boiling tube to be heated by 20 degrees?
2  How long did it take the water-filled boiling tube to be heated by 20 degrees?
3  Is air a conductor or an insulator? Give a reason for your answer.

The best conductors are metals, whilst the best insulators include plastic, paper, wood and air.
Chapter 4 looks at insulators more closely.

## Convection

Liquids and gases are usually poor conductors of heat. However, heat can move through them by CONVECTION.

## ACTIVITY 2.5  Convection in liquids

Put a beaker on a tripod. Put a crystal of potassium manganate (VII) next to the wall of the beaker – see diagram. Carefully fill the beaker with water, trying not to move the crystal. Gently heat the beaker under the crystal with a small flame from a Bunsen burner.

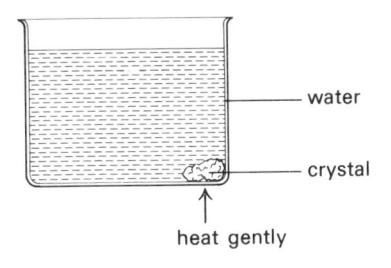

1  Write down what you saw happening.
2  Copy the diagram and add the path of the convection currents.
3  What does this experiment tell you about the density of hot water compared with cold water?

## ACTIVITY 2.6  Convection in air

Put a lighted candle in one half of a beaker. Hang a piece of card in the beaker. Light a drinking straw. After a while smoke will come out of the unlit end. Hold the smoking end of the straw in the empty half of the beaker. Then hold the smoking end of the straw above the candle.

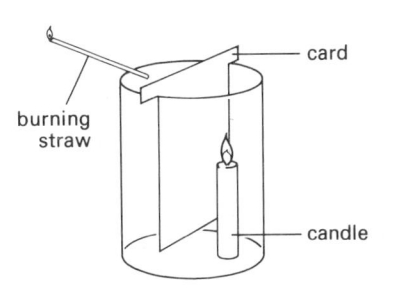

1  What happened to the smoke in the empty half of the beaker?
2  What happened to the smoke above the candle?
3  What was the air doing in the beaker?
4  Copy the diagram and draw arrows to show how the air is moving when the candle is lit.

Convection currents are very important in rooms that are heated by an open fire. Fresh air to make the fire burn properly is drawn into the room from windows and under doors as hot air rises up from the fire. Gas fires and boilers *must* have a supply of fresh air to allow poisonous fumes from the burning gas to escape.

The arrows represent convection currents in a room

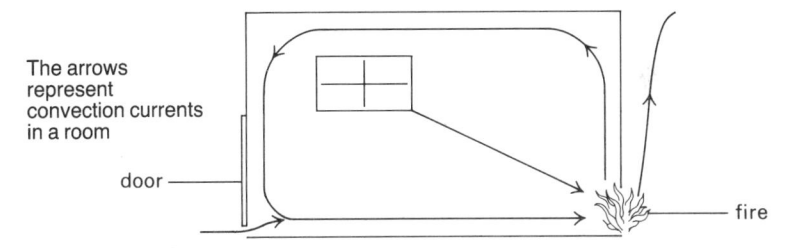

Convection currents in the air can be used for flying. Gliders are lifted up by air currents. If the ground is warm it will heat the air causing it to rise and taking the glider with it. These rising hot air currents are called THERMALS.

Hot air can also be used to lift balloons. When the balloon is filled with hot air it will lift off.

Sea breezes are caused by the land getting hotter than the sea during the day. The air above the land rises as it is warmer, and the breeze we feel is the air moving in from the sea, and then rising over the warmer land. At night the land cools more quickly than the sea and the opposite happens because the air above the sea is warmer than the air above the land.

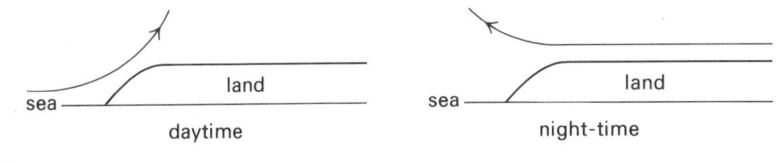

# Radiation

In conduction or convection the heat needs something to move in, like copper or water or air. Heat reaches us from the Sun, however, and there is nothing between the Earth and the Sun. It reaches us by RADIATION which can travel through a vacuum.

### ACTIVITY 2.7   Heat radiation from an electric lamp
Make sure the electric lamp is off, and that the bulb is cold. Hold your hand close to the bulb and turn it on. When you feel the heat, switch it off. Now touch the bulb.
1   Was the glass hot after you had switched off the light?
2   You felt some heat when the light was on. How can you tell that the heat was not conducted or convected to the glass bulb from the hot filament?
3   How *did* the heat get to your hand?

All the heat from the Sun must be radiated. It cannot be conducted or convected because there is nothing between the Sun and the Earth. This heat radiation travels about 150 000 000 kilometres (90 000 000 miles) before it reaches the Earth.

The next experiment is about heat radiation from hot objects, and shows how the radiation is affected by surfaces.

### ACTIVITY 2.8   Emission of radiation
Use a Bunsen burner to heat a thick copper sheet which is blackened on one side. Take the Bunsen burner away and put your hands one on each side of the sheet close to but *not* touching it.
What do you notice about the heat from each side of the sheet?

### ACTIVITY 2.9   Absorption of radiation
Put a shield with a hole in it about 20 cm from an electric heater. Put your knuckle to the hole until it begins to feel warm. Now put a piece of shiny aluminium in front of the hole and touch it with your knuckle.

Now put a piece of blackened foil in front of the hole, with the black side towards the hole. Touch the other side, with your knuckle.

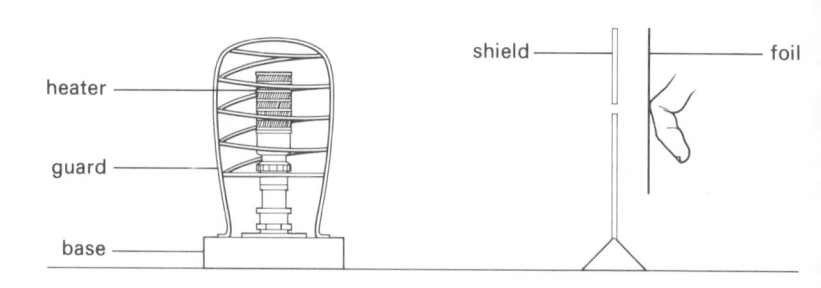

1   What did you feel with the shiny foil in front of the hole?
2   What did you feel with the blackened side towards the heater?
3   Which surface, shiny or black, absorbs heat radiation best? Give a reason for your answer.

In the first radiation experiment the black side of the sheet felt hotter because black surfaces radiate heat better than shiny surfaces.

In the second experiment the black foil got hotter because it takes in heat radiation better than the shiny surface – we say it ABSORBS heat. The shiny foil reflected it so that your hand did not get hot.

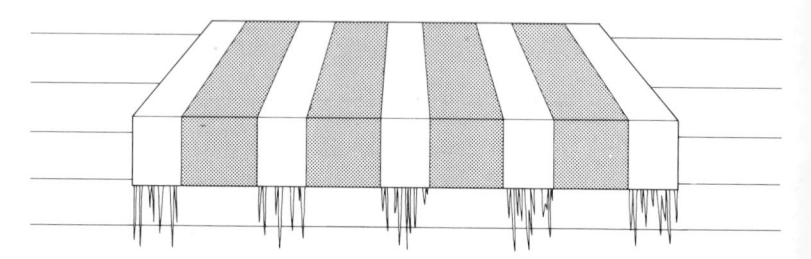

Look at the drawing and explain why there are icicles on the white strips, but none on the black ones.

## Infra-red radiation

Heat radiation is properly called INFRA-RED RADIATION. All objects, including yourself, give off infra-red radiation.

These photographs both show infra-red radiation being given off. The shading in each depends on the amount of radiation, which varies with the temperature of the object.

Doctors can use infra-red photographs called THERMOGRAPHS, to look for ill or diseased parts of the body, which will show up as different colours on the thermograph.

A thermograph

Infra-red radiation can penetrate fog and haze and can also be detected in darkness. Photographs can therefore be taken in these conditions using film which is sensitive to infra-red radiation.

*BBC Copyright photograph*

Foxes photographed at night by an infra-red camera

A night sight used on a rifle or tank gun detects the infra-red radiation given off by the target.

Infra-red lamps are used to ease muscle aches and pains because the radiation goes deep into the body.

Cars are dried under infra-red radiation after they have been painted.

# CHAPTER 3
# Heat and expansion

Expansion joint in a motorway flyover

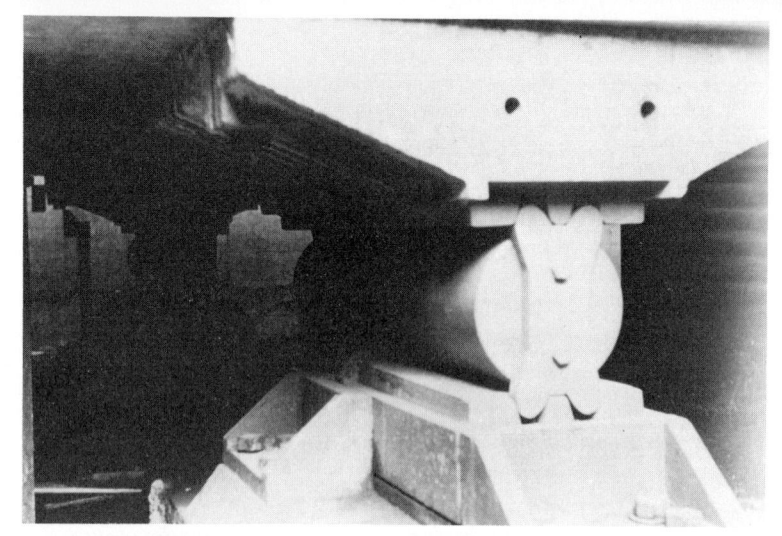

The roller bearings that support each column of the flyover

Expansion loop in an oil pipeline

These photographs have one thing in common. They show that heat is making something expand. Sometimes this is useful, and at other times it is not. The next three experiments show expansion in solids, liquids and gases.

### ACTIVITY 3.1 Expansion of a metal
Set up the metal rod as in the diagram. Heat the rod strongly and watch what happens to the straw. Let the rod cool.

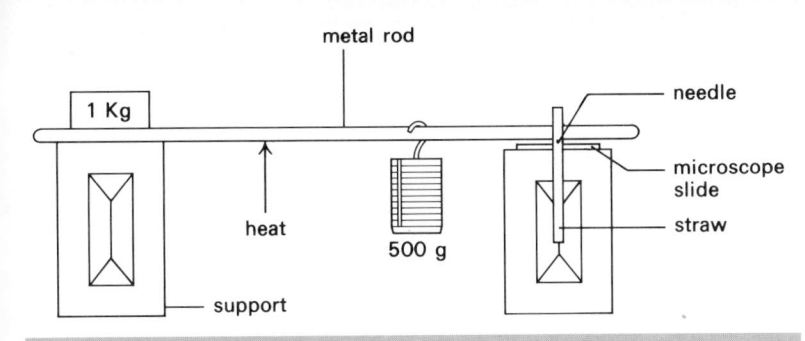

1. What happened to the straw when you heated the rod?
2. Why did this happen?
3. What happened when the rod cooled?

Most solids will expand when heated. A thick glass might shatter if hot water is poured into it. This is because the inside of the glass expands faster than the outside of the glass (see the bimetallic strip). The forces become so great that the glass breaks.

### ACTIVITY 3.2   Expansion of a liquid
Completely fill a flask with water. Carefully insert the bung and tube so that the water rises up the tube, as in the diagram. Put the flask into the water bath and heat the water. Do not let it boil. Watch the water level in the tube.
1. What happened to the water in the flask when it was heated?
2. Why did it do this?

### ACTIVITY 3.3   Expansion of a gas
Put the end of the tube into the water in the beaker, as in the diagram. Warm the air in the flask by holding it in both hands. Watch what happens. Cool the flask with a cloth soaked in cold water, or with a freezer spray.
1. What happened when the air was heated by your hand?
2. Why did it do this?
3. What happened when the air was cooled?
4. Why did it do this?

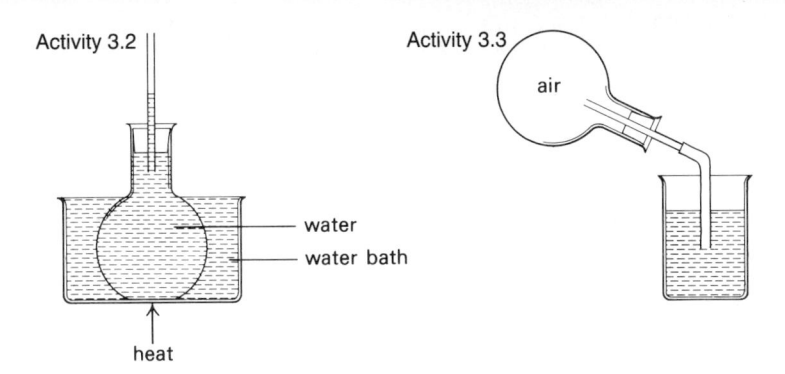

### The unusual behaviour of ice
You have seen that solids and liquids contract when cooled, but water does not always do this. When the temperature reaches 4°C water stops contracting and starts to expand again. When it freezes solid its volume has increased by about one-tenth. It is this increase in volume which cracks water pipes in winter when the water freezes. However, you do not find out about it until the ice thaws! The increase in volume when ice is formed means that ice is less dense than water and this is why ice floats on water. This is important for things living in the water as the ice then acts to insulate the water below and stops it from freezing further.

Ice will melt at a temperature lower than 0°C if pressure is put on it and it will freeze again when the pressure is removed. This is called REGELATION. In the diagram below, the wire passes right through the ice block, but the block is not cut in half — it is still a complete block.

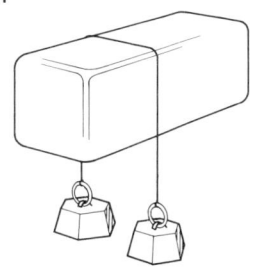

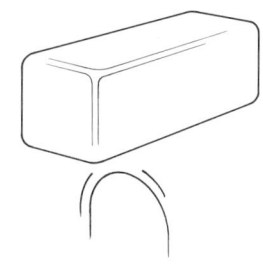

Under the wire the pressure is high, so that the freezing point of the ice is less than 0°C. This means that the ice under the wire melts so the wire moves down. After the wire has moved down the pressure above it is reduced, so the water again freezes at 0°C.

This effect enables ice skates to slide over ice. The pressure of the skates melts the ice immediately beneath them and a thin film of water is formed which enables the skates to move freely. If the ice did not melt they would not move. Snowballs hold together because you melt some of the snow by squeezing it in your hands. If the snow is very cold you may not be able to exert enough pressure to melt the snow and the snowball will not hold together. The experiments above show that solids, liquids and gases will expand when heated, and contract when cooled. This effect is used to control temperature. A device which controls temperature is called a THERMOSTAT.

A thermostat for controlling room temperatures

## How thermostats work

A common type of thermostat uses a BIMETALLIC STRIP.

A bimetallic strip is made by joining two strips of different metals together. The commonest type of bimetallic strip is made of brass and invar. Invar is an alloy of iron and nickel, and it does not expand very much when heated.

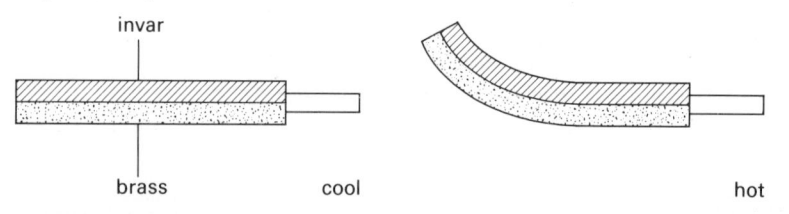

When the strip is heated the brass expands more than the invar and the strip bends. Which metal is on the inside of the bend?

**ACTIVITY 3.4 How a bimetallic strip controls an electric circuit**

Set up the apparatus as in the diagram. Leave a gap of about 1 cm between the nail and the bimetallic strip. Heat the strip with a Bunsen flame.

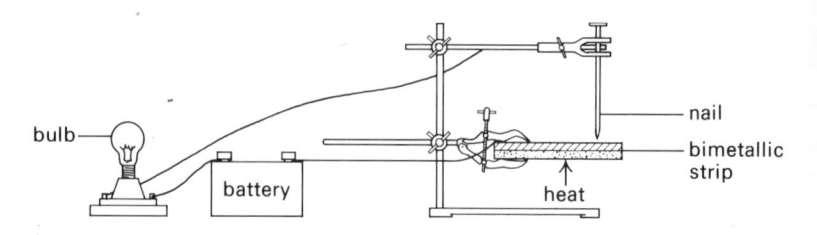

1. What happened as the strip was heated?
2. What happened when the strip cooled?
3. What side was the invar on – the top or the bottom?
4. What useful job could this circuit be used for?
5. How would you adjust the apparatus to get the circuit to switch on for (a) a higher temperature and (b) a lower temperature?

### Rod thermostats

A rod thermostat does not use a bimetallic strip, and is commonly used in ovens and hot-water tanks.

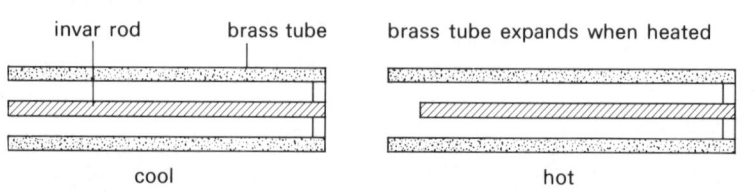

When the thermostat is cool the ends of the rod and tube are level. When the thermostat warms up the rod and tube expand, but by different amounts.

## ACTIVITY 3.5  How the oven thermostat works
Here is a diagram of the rod thermostat in a gas cooker oven. Look at it, and then copy out the passage below, filling in the missing words from the list alongside.

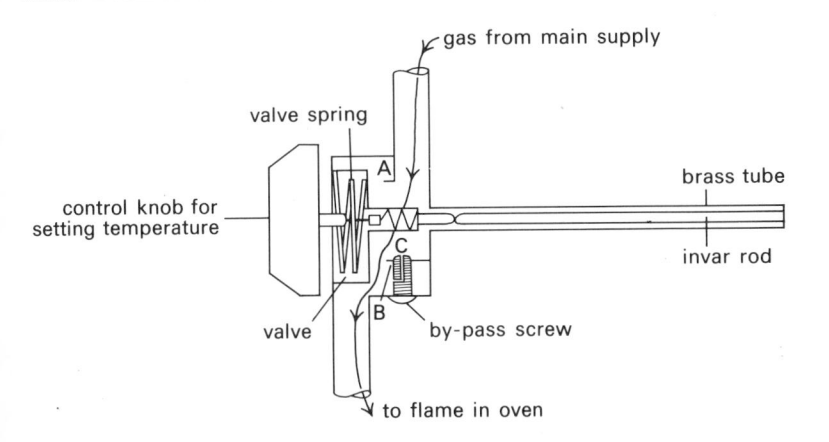

The rod thermostat will keep the ...... of the oven steady. Gas flows in from the ...... ......, past the ...... ...... and out to the ...... in the oven. If the valve spring pushes the valve to the right it closes onto A and B and reduces the flow of the ......, so the flame goes down. Some gas however will flow through C to make sure the flame does not ...... ......
As the oven gets hotter the tube and the rod ...... As brass expands more than invar the tube ...... the rod away from the valve. The valve spring then ...... the valve onto A and B, and the gas flow is ......
When the oven cools the brass tube ...... and pushes the rod against the valve. This lets the gas flow again and the flame gets ......

by-pass screw
hotter
reduced
go out
gas
temperature
pulls
contracts
main supply
flame
expand
pushes

## ACTIVITY 3.6  How a hot water thermostat works
Copy out this passage and after looking at the diagram below fill in the missing words from the list given alongside.
When the water is cold the rod and tube are the same ...... The mercury in the switch is ......, and covers the contacts A and B. The electric current flows through the mercury to the water heater.
When the water is hot the brass tube ...... and ...... C down, so that the mercury does not cover the contact B any more. This breaks the ...... and the heater goes ......

circuit
expands
off
level
pushes
length

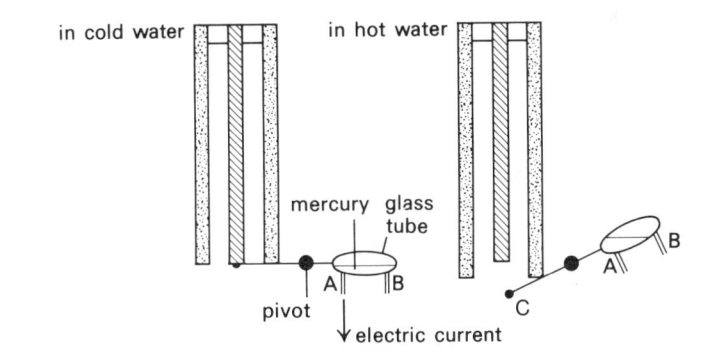

## ACTIVITY 3.7  Heat and expansion
Write down the answers to these questions.
1  What are thermostats used for?
2  Make a list of where thermostats are used in your house.
3  Look at the photographs on page 12.
   a  Why was the flyover on rollers?
   b  Why did the oil pipeline have a loop in it?
4  Why are metallic tape measures made out of invar?
5  When telegraph wires are hung, why are they hung loosely, and not pulled tight?

# CHAPTER 4
# Heat and insulation in the house

It is very important that you know how to stop your house losing heat. Firstly, you save money if you use less heat. Secondly, if everybody in the country uses less heat then supplies of gas, coal and oil will last longer. You need to know how heat is lost and how you can reduce the loss.

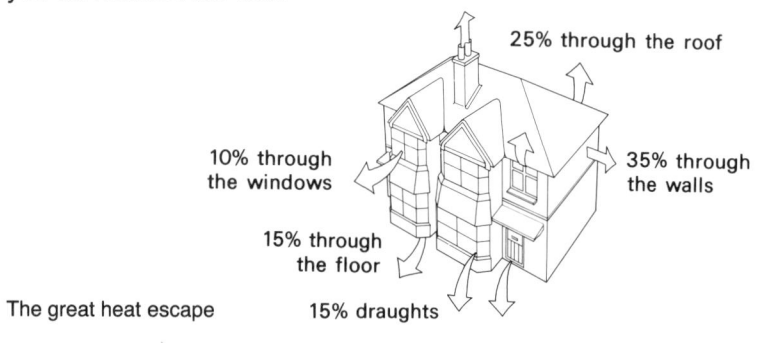

25% through the roof
10% through the windows
35% through the walls
15% through the floor
15% draughts

The great heat escape

### Floors
A floor will lose less heat if it has a thick carpet covering it. (This also cuts down noise.) Draughts can blow up between the floorboards, and carpet underlay stops this. Newspapers can also be used.

### Roofs
On a frosty or snowy day you can often see which houses have insulation in their lofts. The frost or snow will last longer on the insulated roof because the heat cannot get through to melt it. The commonest way of insulating a loft is to use GLASS WOOL or MINERAL WOOL FIBRE rolls.

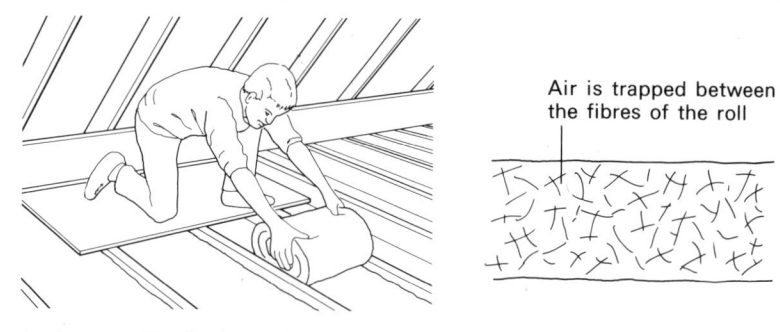

Laying glass fibre loft insulation

Air is trapped between the fibres of the roll

The roll is made of many thin fibres of glass or mineral wool. Air is trapped in the spaces and acts as an insulating layer.
The fibres can break off and make your skin itch, so use gloves when you handle it. The roll has to be at least 75 mm (3 inches) thick to insulate well, and 100 mm (4 inches) is reckoned to be the best thickness. It is not worth putting any more on.
LOOSE-FILLING is the other main type of roof insulation. Little bits of polystyrene, or a material called vermiculite are put between the joists. Once again air is trapped for insulation. The advantage of loose-filling is that you just pour it out of the bag and you do not need to do any cutting or fitting to fill awkward spaces. Less popular methods of insulation are to lay rigid polystyrene sheets on the joists, or put down aluminium foil. The foil reflects heat back into the house, so it works differently from the other methods.

### Walls
Many houses are built with two layers of bricks which have a gap or cavity of about 50 mm between them. This is called a CAVITY WALL. It prevents damp getting to the inside wall and also insulates the house. If the gap is filled with plastic foam or mineral wool blown into it, then the heat loss is cut down even more, because the foam is a better insulator than air, and it reduces convection currents.

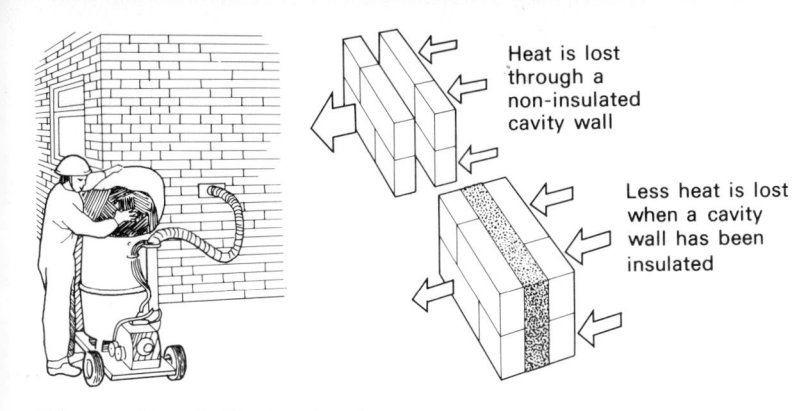

Filling a cavity wall with mineral wool

## Doors and windows

A draught is caused by cold air blowing into a warm room. It can do this at doors, windows or between the floor boards. The best way to stop draughts at doors is to fit a draught excluder. There are different kinds, but they are all cheap and easy to fit.

Different kinds of draught excluders for doors

Windows are best insulated by DOUBLE GLAZING.
Double glazing means putting another sheet of glass or clear plastic onto the window. This forms a layer of air, which acts as an insulator. Unfortunately double glazing is an expensive way of insulating. Thick curtains, when closed, will also reduce heat loss.

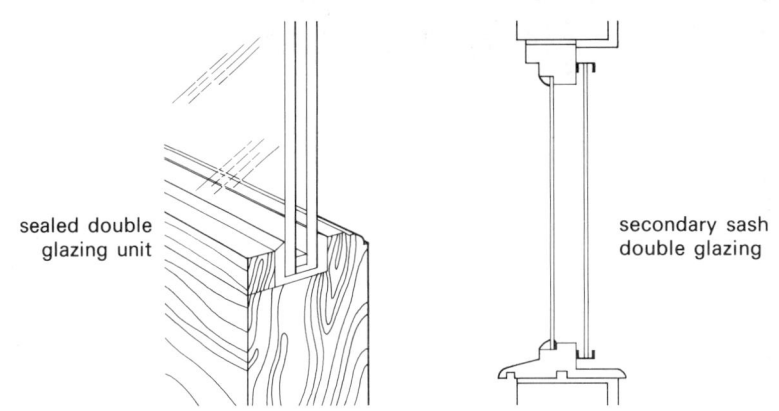

## Chimney

If the chimney is not used any more, it should be blocked off, but a gap must be left for ventilation, especially if a gas fire is used. This is because the fire will burn up too much of the oxygen in the room, if fresh air cannot get in, and produce poisonous fumes.

## Hot water tanks and water pipes

To reduce the amount of heat being lost from hot water tanks and pipes, a thick jacket of insulating material is wrapped around them. This is called LAGGING. Cold water pipes also need lagging, but this is to stop them freezing in winter and possibly bursting. Lagging reduces heat loss by about four-fifths.

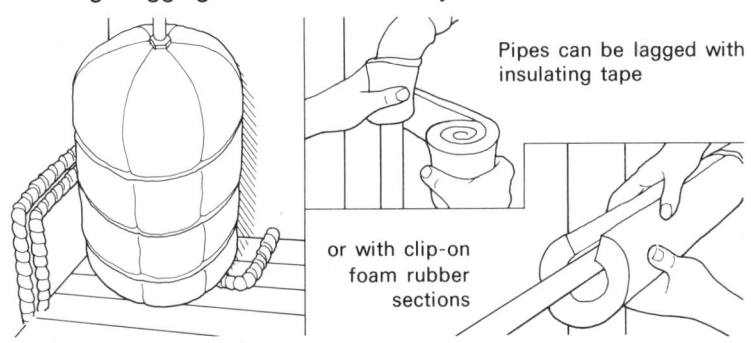

Well-lagged hot water tanks and pipes

# CHAPTER 5  Clothes

The clothes you wear are very important in keeping you warm or cool. Look at these two pictures.

The Eskimo boy wears thick furs to keep warm, and they cover all of his body. But why do the men from NW Africa wear long robes which also cover their bodies? In hot weather you wear as little as you can, so why do people in hot lands often wear long clothes? The reason is not clear, and could be any of these.

(a) The long robe traps a layer of air, and as you saw in Chapter 2 a layer of air is a good insulator. This air insulation therefore helps to stop the Sun's heat getting to the body and keeps the body cool.
(b) The robes are often light in colour and so they reflect some of the Sun's rays.
(c) The robe stops the sweat evaporating quickly and stops the body losing too much moisture.
(d) In some countries the long robes are worn for religious reasons.

The Eskimo's clothes are simpler to understand. Under his furs he wears two or three thin shirts. Air is trapped between each one and stops his body heat getting out. Air is also trapped in the fur itself, and this is why animals in cold countries usually have fur.

The feathers of birds do the same thing, and in winter you often see birds fluffing up their feathers to trap a layer of air to keep warm.

A bird fluffing its feathers to keep warm

Goose-pimples happen when you are cold because you are trying to raise the hairs on your body to keep warm – although you do not have very many!

Modern outdoor clothes like anoraks or car jackets work in the same way. They are made of materials which have lots of air pockets in them. The air is trapped in them and stops the body heat getting out. Mountaineering clothes are lighter than they look because the jacket, leggings and sleeping bag are made of pockets of air trapped by very light materials. The rescuers in the photograph on page 5 are wearing clothes like this.

You may sleep in a bed with a DUVET or CONTINENTAL QUILT. This keeps you warm in the same way as the mountaineering clothes.

### ACTIVITY 5.1  Which materials keep you warmest?

Some materials are better than others for keeping you warm. This experiment will help you find out which are the best. You can use cotton, wool, nylon and linen from bits of old clothes like shirts, trousers, skirts and jackets.

Each material you are testing is wrapped around the beaker.

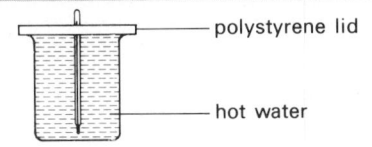

Fill the beaker with boiling water at the start of each experiment and put the polystyrene lid on. Leave the beaker for an hour, taking the temperature every 15 minutes.

Copy the table and enter your results, heading each column with a different material.

| Time in minutes | Temperature in °C |   |   |
|---|---|---|---|
| 0 | | | |
| 15 | | | |
| 30 | | | |
| 45 | | | |
| 60 | | | |

1  Which material do you think is the best for keeping you warm? (Give a reason for your answer.)
2  What clothes are made out of this material?
3  Which material was the worst for keeping heat in?
4  What clothes are made out of this material?

### ACTIVITY 5.2  Clothes

Write down the answers to these questions.

1  Look at your own clothes. Which are best for keeping you (a) warm and (b) cool? Explain why they are good.
2  You are going walking on a cold day. Which clothes from this list would you wear: jeans; corduroy trousers; tee-shirt; anorak; denim jacket; plastic mac? Give your reasons.
3  Do girls wear tights to keep their legs warm?
4  Spacemen wear two or three thin vests to keep warm. Why does this work?
5  A space blanket is a thin sheet of plastic which has a shiny surface on one side. It is carried by mountaineers for emergency use if they are caught in very bad weather. They wrap themselves up in it with the shiny side next to their clothes. Explain how the space blanket helps to keep them warm.

## CHAPTER 6

# Keeping food hot or cold

For people going camping, cycling, walking, etc., keeping food hot or cold after it has been cooked is very important. The main methods are the VACUUM or 'THERMOS' flask, and the COOLER BOX. Each can be used for keeping food either hot or cold.

## The vacuum flask

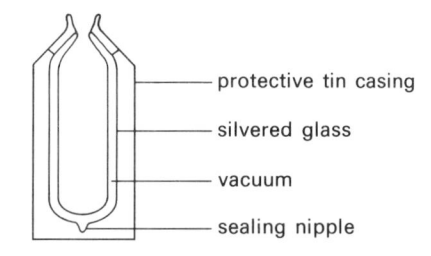

A vacuum flask looks like this. It is easy to take one to pieces and see the main part, the silvered inner flask. This is made of thin glass which is silvered like a mirror. Notice that it is not a single thickness, but is two walls with a space in between. Air is removed from this space and the glass is sealed at the sealing nipple. This leaves a vacuum which reduces heat loss. The silver sides also reduce heat loss by reflecting heat back. The flask is put in a metal or plastic casing to protect it.

### ACTIVITY 6.1
You will have two vacuum flasks, one of which has had the sealing nipple broken off. Fill both with boiling water. After 10 minutes feel the outside of each.

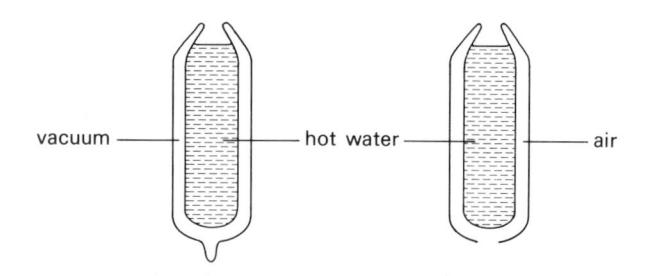

1   Was there a temperature difference between the broken flask and the unbroken flask before the water was put in?
2   What difference was there between the flasks when you touched them after 10 minutes? Give a reason for the difference.
3   In which flask would ice melt first, and why?
4   Imagine that you have two unbroken vacuum flasks, but one is made from clear glass, i.e. it is not silvered. You fill both flasks with boiling water. What difference would you feel between the flasks after 10 minutes? Give a reason for your answer.

## Cooler and freezer boxes

It might seem funny to say that the cooler box also keeps food warm, but like the vacuum flask it reduces the amount of heat getting in or out.

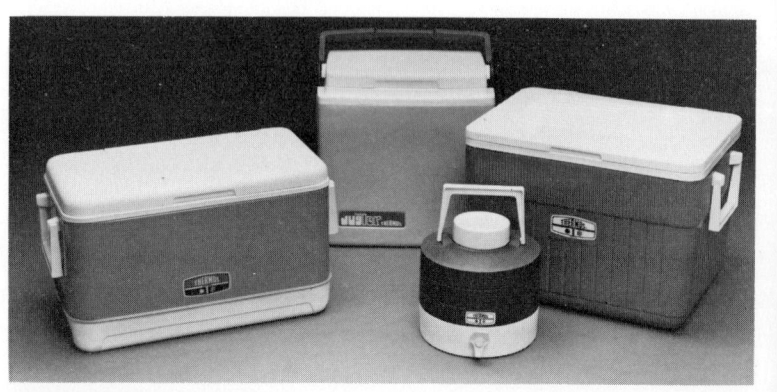

Cooler and freezer boxes

## ACTIVITY 6.2  Foam insulation

Cut a polystyrene tile to make a small box for an ice cube. The ice cube is put into the box, which is held together by elastic bands. Leave another similar ice cube nearby.

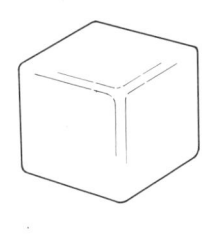

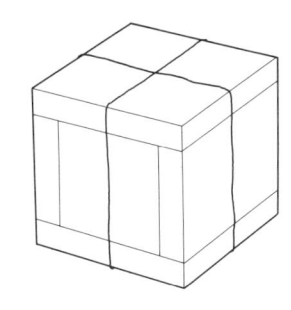

1. Which ice cube melted first?
2. How does this experiment show that polystyrene foam is a good insulator?
3. Why is polystyrene foam a good insulator?

A cooler box can be used to cook food like stews if the food is heated up first. It is then put into the box and left for the normal cooking time. The food stays hot and so cooks. Any box filled with a lot of insulating material like blankets, crumpled up newspapers or straw can be used to cook a pot of stew in this way.

## The refrigerator

In Chapter 1 you saw that when a liquid evaporates it cools down its surroundings by taking heat from them. This is the main idea behind how the fridge works, but first you need to know something about boiling and pressure.

The commonest way of evaporating a liquid is by heating it until it boils. Water boils at 100°C and changes to steam; but you can make water boil at a temperature lower than this. This experiment shows how.

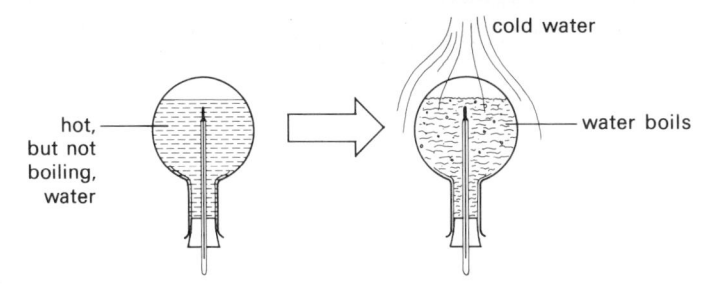

Cold water is poured onto a flask of water which is very hot but not boiling. The hot water boils because the cold water makes the hot *air* above the water contract (see Activity 3.3). This reduces the pressure above the hot water and it boils.

This experiment shows that reducing the air pressure makes water boil at a lower temperature. This is why tea, which needs water at 100°C cannot be made satisfactorily up a high mountain unless the water is boiled under pressure. The air pressure is less at the top of a mountain, so the water boils at less than 100°C and will not properly brew tea.

The refrigerator works on two principles. When a liquid evaporates it takes heat from its surroundings, and if the pressure on a liquid is reduced it will evaporate.

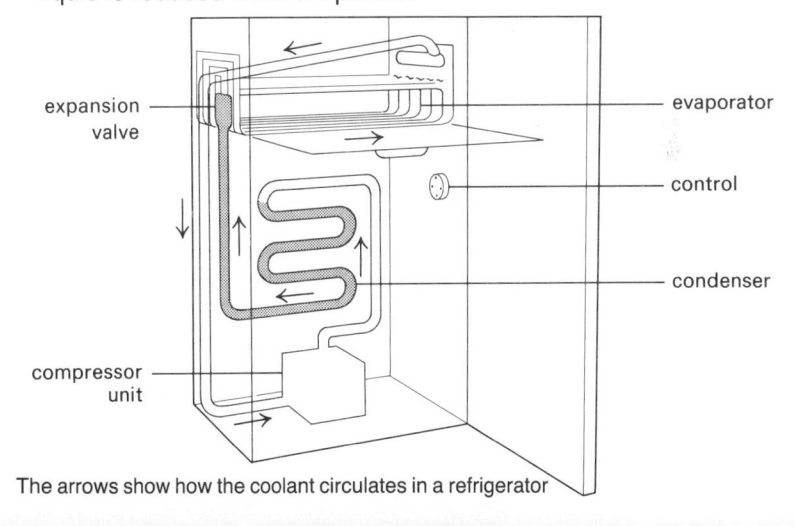

The arrows show how the coolant circulates in a refrigerator

Look inside a refrigerator. The box at the top, known as the freezer compartment, is where the liquid evaporates and cools its surroundings. You can usually see the tubes in which it runs, and you must be very careful not to puncture them when cleaning the refrigerator.

After the liquid has evaporated it goes back to a COMPRESSOR which increases the pressure on the vapour and turns it back to a liquid again. This is called CONDENSING. During the condensing heat is released from the vapour. You can feel this heat being released if you put your hand over the back of the refrigerator when it is running. *Do not touch anything.* There should be a clear space behind a refrigerator so that the heat can escape. This is the heat which is taken from the food in the refrigerator.

### Heat pump

The refrigerator takes heat out of the food in the refrigerator box and pumps it into the air. If the refrigerator is run backwards it will take heat from the air and heat up the refrigerator box. This is the principle of the heat pump, which is how some buildings are heated. A supply of energy is needed only for the electric pump. The heat energy is taken from the air or from a nearby river. Some dairies use heat pumps to warm up their offices. The heat is taken from the milk when it is cooled and pumped into the building.

## Pressure cooker

You have seen that reducing the pressure on water makes it boil at a lower temperature. What do you think happens if the pressure is increased? Yes, the water boils at a *higher* temperature. This is what makes a pressure cooker useful for cooking.

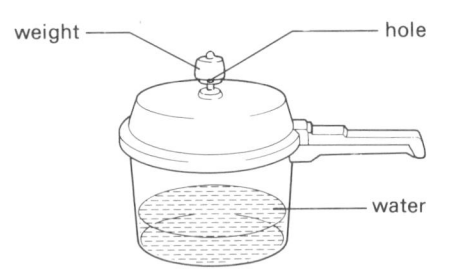

### ACTIVITY 6.3  How the pressure cooker works

Copy the passage and put in the missing words from the list alongside.

When the water boils it forms …… which …… the pressure in the cooker. As the pressure gets higher the …… of the boiling water also gets higher. The pressure cannot continue to go up and up or else the cooker will ……, so a safety device is needed. The …… at the top is a weight placed over a small hole. When the steam pressure gets too high the weight is ……, and the steam escapes.

The …… temperature allows the food to be cooked faster.

temperature
higher
valve
lifted
raises
explode
steam

# CHAPTER 7

# Temperature and heat energy

| Temperature in °C | |
|---|---|
| 20 000 000 | Interior of the Sun |
| 5 000 | Temperature of surface of the Sun |
| 3 650 | Tungsten melts. It is the metal with the highest melting point |
| 3 500 | Carbon arc flame |
| 2 500 | Filament in a light bulb |
| 1 769 | Platinum melts |
| 1 550 | Hot Bunsen flame |
| 1 400 | Steel melts |
| 1 084 | Copper melts |
| 1 063 | Gold melts |
| 200 | Fat in a chip pan |
| 100 | Water boils |
| 58 | The hottest air temperature yet recorded (Libya, 1922) |
| 50 | Hot bath |
| 37 | Body temperature |
| 25 | Hot day |
| 10 | Cool day |
| 0 | Water freezes |
| −39 | Mercury freezes |
| −78 | Dry ice – solid carbon dioxide |
| −88 | Lowest air temperature ever recorded (Antartica, 1960) |
| −95 | Methanol (methyl alcohol) freezes |
| −196 | Nitrogen gas becomes liquid |
| −270 | Temperature of outer space |
| −273 | Lowest temperature that can ever be reached – absolute zero |

## Temperature-measuring instruments

All the temperatures in the chart above have had to be measured in some way. You may imagine that quite a variety of temperature measuring instruments would have been used.

*Mercury and alcohol thermometers*

### ACTIVITY 7.1
We usually use a MERCURY thermometer to measure how hot something is. Look at a thermometer and find the following parts: the mercury bulb; the mercury thread; the scale; the capillary.

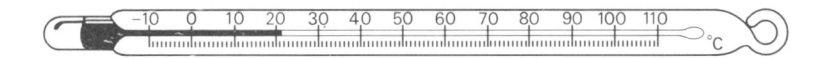

1. Draw a diagram of a thermometer and label these parts.
2. What happens to the mercury if you touch the bulb?
3. Why does it do this?
4. What happens when you stop touching the bulb?
5. Why does it do this?

The scale is marked in units usually called degrees Centigrade, because the weather forecasters still use them, but the proper name is degrees Celsius, written °C.

As mercury is expensive, cheaper thermometers use alcohol (methanol) which is dyed red to make it more easily seen. There is another reason for using an alcohol thermometer, look at the chart of temperatures and try to decide what that reason is.

*Clinical thermometer*

A clinical thermometer is used to measure body temperatures.

To read the temperature accurately the scale only goes from 34°C to 42°C (94°F to 108°F), so never use the thermometer for anything else – it could easily break.

The kink in the tube is a constriction (a narrow part) to stop the mercury running back so that you can read the temperature easily. Your teacher will show you how to get the mercury down by shaking the thermometer with a flick of the wrist.

### ACTIVITY 7.2
Copy the diagram of the clinical thermometer and label the bulb, the constriction and the mercury thread. Indicate normal body temperature.

### ACTIVITY 7.3
Measure your own temperature by putting the thermometer under your tongue (*do not bite it!*) and leaving it there for about one minute. The thermometer should be rinsed under *cold* water and sterilised in an antiseptic solution (some Dettol in water) before being used again. You will find that different people have slightly different temperatures. Your own temperature may vary throughout the day and from day to day. Average body temperature is about 37°C (98.4°F). Your own temperature might be a degree above or below this without any ill effects.

## Thermocouples
A thermocouple is used to measure very high temperatures. Two wires of different metals, usually copper and constantan, are twisted together at one end, making what is known as the junction. When this junction is heated a small electric current is produced. The hotter this junction becomes, the bigger the current.

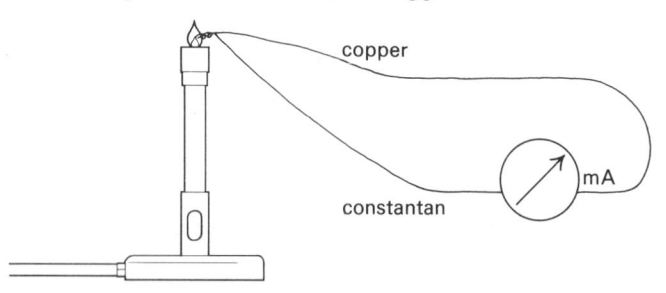

### ACTIVITY 7.4
Make a thermocouple as shown in this diagram. Put the twisted end into boiling water (100°C) and into a Bunsen flame. See if you can work out the approximate temperature of the Bunsen flame. If the flame is hot enough the copper might melt. Find from the temperature chart, the temperature at which copper melts.

## Optical pyrometer
An optical pyrometer is sometimes used when the temperatures of molten metals or very hot flame temperatures have to be measured. (Why could a thermocouple not be used?)

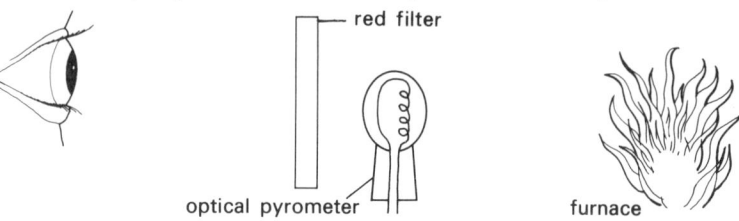

When the colour of the filament in the lamp is the same as the furnace they both have the same temperature. By measuring the current through the filament the temperature can be calculated. The temperature of the surface of the Sun can be measured in this way. DO NOT try to do it yourself. It is dangerous to look directly at the Sun; you may damage your eyes.

## Bimetallic-strip thermometer
Thermometers which are used in freezers or for cooling are often bimetallic strips which have been coiled up.

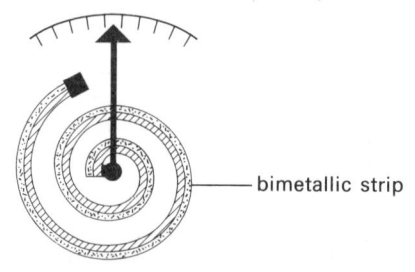

**ACTIVITY 7.5**
Copy the diagram of the bimetallic-strip thermometer and explain how it works.

*Resistance thermometer*
The electrical resistance of a metal depends on its temperature, so by measuring the resistance its temperature can be calculated.

**ACTIVITY 7.6**
Find out whether the resistance of copper increases or decreases when its temperature is raised. Design your own experiment. You will need a 1½ V battery, an ammeter, about two metres of thin (36 or 38 swg) insulated copper wire, a beaker of cold water and a beaker of hot water.

# Heat energy

Heat is a form of energy. This means that you can measure a quantity of heat in joules. 1 kg of water needs 4200 J of heat energy to raise its temperature by 1 Celsius degree, written 1C°. This amount of heat energy is called the SPECIFIC HEAT CAPACITY of water. The specific heat capacity varies for different materials: Paraffin 2130 J; Water 4200 J; Lead 126 J; Copper 385 J; Air 993 J.
The same amount of heat is *lost* if the material *cools* by 1C°.

**ACTIVITY 7.7  To find out how much heat energy is given out when 1 cm³ of methylated spirit is burnt.**

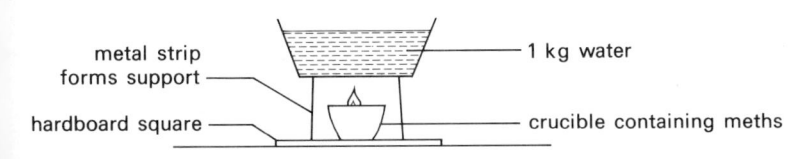

metal strip forms support — 1 kg water
hardboard square — crucible containing meths

Put 1 kg or 0.5 kg of water in the metal container, and put the container on its support. Take the water temperature. Put 2 cm³ of meths into the crucible and put it under the metal container. Light the meths, and while it is burning, gently stir the water. When the meths is burnt out take the highest temperature of the water.
1  Copy out this passage and fill the missing spaces with your results.
The meths heated up my ...... kg of water from ...... °C to ...... °C. This is an increase of ...... C°.
1 kg of water needs 4200 J of heat energy to raise its temperature by 1C°. To raise the temperature of my ...... kg of water by 1C° I therefore would need ...... J. In the experiment my ...... kg of water went up by ...... C° so it must have had ...... J of heat energy from the meths.
I used ...... cm³ of meths which gave out ...... J of heat energy to the water, so 1 cm³ of meths will give ...... J.

The correct answer is 17 953 J.

2  There are two main reasons why your answer is (or should be!) less than 17 953 J. What are they?
3  Why was the water stirred when the meths was burning?

# Melting and boiling

As the temperature of a solid gets higher it first melts and turns to a liquid. The liquid then boils and turns to a gas or vapour. To understand why this happens you need to know that matter is made of small particles called atoms or collections of atoms called molecules.
When any solid melts, its atoms or molecules move faster and faster losing their regular pattern, and the solid turns into a liquid. When the liquid is heated the atoms or molecules move even faster, and by escaping from the liquid they turn it into a gas or vapour.

If a tin of air is heated it will blow the lid off because the gas molecules are moving with even greater energy, and so hit the lid with greater force.

The temperatures at which different solids melt and boil vary a lot as you can see from the chart at the beginning of this chapter.

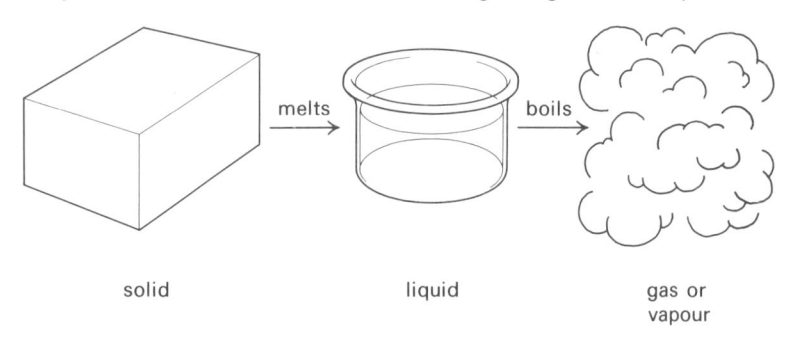

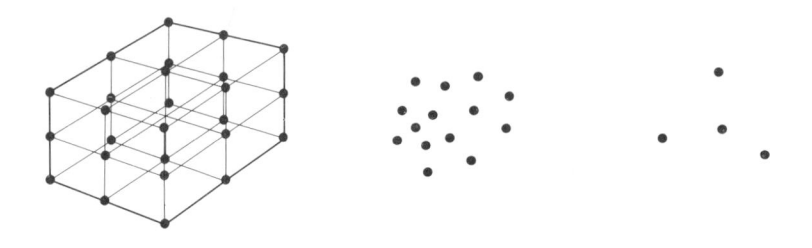

solid — melts → liquid — boils → gas or vapour

atoms or molecules are regularly arranged

atoms or molecules are randomly arranged

atoms or molecules are far apart

### ACTIVITY 7.8 The molecular model of heating
Look back at Activities 1.2 (Keeping cool), 2.1 (Heat being conducted), 3.1, 3.2 and 3.3 (Expansion of a metal, a liquid and a gas).

For each activity write an explanation of how the effect shown could be explained by the movement of molecules or atoms.

### Absolute zero
If any gas is cooled its volume gets less and less. This is shown in this graph.

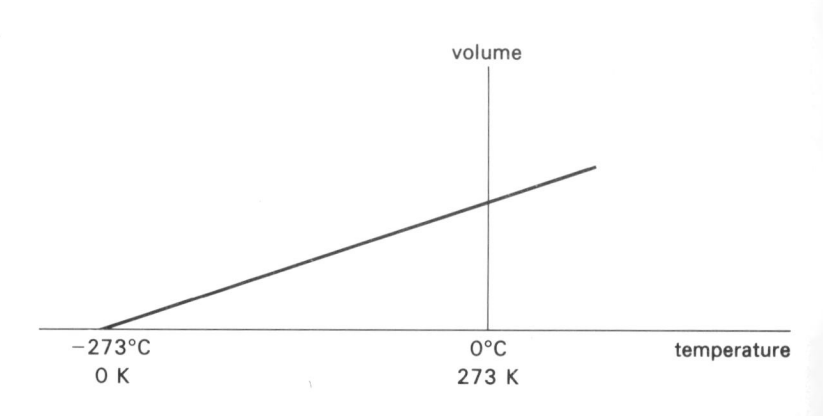

−273°C / 0 K     0°C / 273 K     temperature

At −273°C the gas is thought to have no volume because we imagine that the gas molecules have stopped moving, and so cannot lose any more energy as heat. This, and other ideas, are why scientists think that the lowest temperature that can be reached is −273°C, called ABSOLUTE ZERO. A different temperature scale starts at absolute zero and has units called degrees Kelvin, written as K, e.g. 185 K. This means that the temperature at which water freezes is 0°C on the Celsius scale and 273 K on the absolute zero temperature scale, and it would boil at 373 K. This scale is used by scientists, and is not likely to replace the Celsius scale for everyday use.

## OPTION A

# How do we get heat?

## Burning
Burning a fuel gives us heat for our homes.

Gas fire

Paraffin fire

Solid fuel fire

### ACTIVITY A.1
1  Make a list of all the fuels used today which are burnt for heat.
2  Which fuels do you think were not used 100 years ago?
3  Which fuels do you think were not used 1000 years ago?

## Electricity
Electricity gives you heat, but you do not burn it.

### ACTIVITY A.2   Heat from electricity
Connect a piece of wire wool or very thin wire across the terminals of a battery.
1  What happened when the electric current flowed through the thin wire?
An electric current can heat up a metal wire without melting it as in an electric fire.
2  Make a list of all the places and things in your home which use electricity for heating. For each one on your list say if you could use something else instead of electricity.

## Friction
Rub your hands together. They get warm because there is friction between them. When two surfaces rub together there is friction, and if there is enough it causes heating.

### ACTIVITY A.3
1  Make a list of all the situations that you know of where friction causes heat.
2  Where is heating by friction undesirable?
3  How would you try to reduce friction?
4  Where is heating by friction useful?

Friction welding can join two bits of metal together. One piece of metal is rotated against the other until a high temperature is reached.

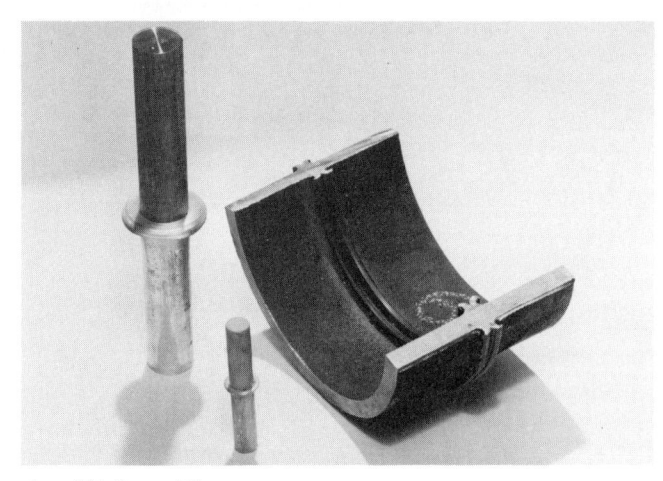

Examples of friction welding

The rotating is stopped when the two pieces melt at the surfaces and are welded together under a high pressure.

## Chemical reactions

> **ACTIVITY A.4  Heat from chemicals**
> Add a few drops of water to a spatulaful of dehydrated copper sulphate in a test tube. Feel the test tube.
> What happened when the water was added to the dehydrated copper sulphate?

An explosion may occur when chemicals react so fast that a lot of heat is given off in a very short time. Matches or explosive caps for toy guns are examples of small explosions that turn chemical energy into heat energy. What other forms of energy are involved? Burning is also a chemical reaction.

## Sun

The Sun's radiation heats the Earth and this energy is used in solar furnaces which focus the radiation.

A 'solar stove'

> **ACTIVITY A.5**
> 1   Make a list of all the uses of the Sun's radiation.
> 2   When is the Sun's heat a 'nuisance'?

## Radioactivity

A piece of radioactive material, for example radium or plutonium, will become hot due to the atomic reactions going on inside it.

> **ACTIVITY A.6**
> Find out what the atomic reaction is and how it makes heat. This will be difficult to understand, so ask your teacher for help, or read about it in the library.

OPTION B

# Fuels for home heating

## ACTIVITY B.1  Fuels in the home
1  List the fuels: oil, coal, gas, electricity (classed as a secondary fuel), used in your house and say what each is used for (cooking, lighting, heating, heating water, etc.)
2  Can the different uses use different fuels? For example, could you light your house without using electricity? Explain how you could do it and do the same thing for the other uses in your list.

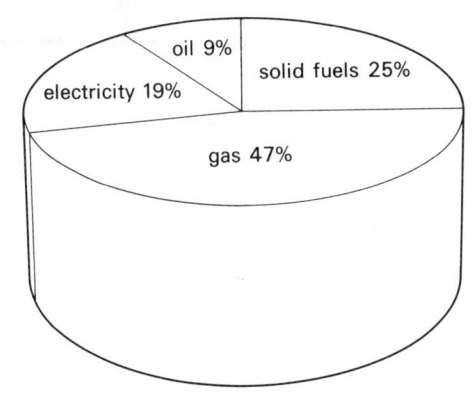

Fuel used in heating the home in the UK in 1978

- oil 9%
- solid fuels 25%
- electricity 19%
- gas 47%

## ACTIVITY B.2  Home heating
1  Conduct your own survey of the heating which people use and make a pie diagram like the one opposite for your results. 2  In what ways are your results different from the UK figures? If your results are different try to explain why.

## ACTIVITY B.3  Heating costs
This table gives the average costs of heating a three-bedroomed house in 1980.

| Method | Cost per year |
| --- | --- |
| Oil central heating | £450 |
| Gas central heating | £200 |
| Electricity | £770 |
| Coal | £600 |
| Paraffin | £450 |
| Bottled gas | £700 |

1  What is the cheapest way to heat a house?
2  Why is it the cheapest?
3  Do you think it will always be the cheapest? Why?
4  What is the dearest form of heating?
5  Why is it so dear?
6  Is it likely to stay the dearest? Why?
7  What are the advantages of central heating?
8  For each of these central heating fuels say what their advantages are: oil; gas; electricity; solid fuels.

**OPTION C**

# Heat saving in your home

### ACTIVITY C.1
Draw a picture of your home, or make a model of it.
On it show how and where heat is lost in your home and how you try to keep it in. Chapter 4 on how houses lose heat will help you.

### ACTIVITY C.2
Conduct a survey of your class and school to see how their homes are insulated. Ask about the four types of insulation: loft insulation, double glazing, draughtproofing, and cavity wall insulation.
Make a chart or poster of your results.

### ACTIVITY C.3
Here are posters asking you to save heat. Have a go at drawing your own.

### ACTIVITY C.4
You can compare different insulating materials in the same way as you compared clothes in Activity 5.1. Try to compare as many as possible of the methods described in Chapter 4.

### ACTIVITY C.5
Make a collection of insulating and draughtproofing materials. You can then label them and put them out for display, with posters and pictures showing how they are used.

OPTION D

# Home and hot water heating

In a gas, oil or solid fuel boiler the fuel is burnt and the flame is played over a HEAT EXCHANGER. This is a metal compartment with fins, through which the cold water passes.

A heat exchanger

In an electric heater there is no flame. The electric current flows in the coil of wire known as the HEATING ELEMENT and heats it up. The element cannot be put directly into the water, so it is surrounded in a ceramic insulator, and then put in a waterproof metal case.

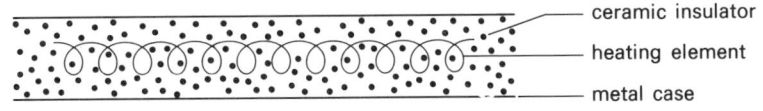

## Heating systems

There are two main systems for circulating hot water around a house. They are GRAVITY systems and PUMPED systems. In a gravity system hot water rises from the boiler because of convection (see Activity 2.5). After it has lost its heat in the hot water tank or radiator it is more dense and now falls, due to gravity, back down to the boiler again. Follow the water's path on the diagram.

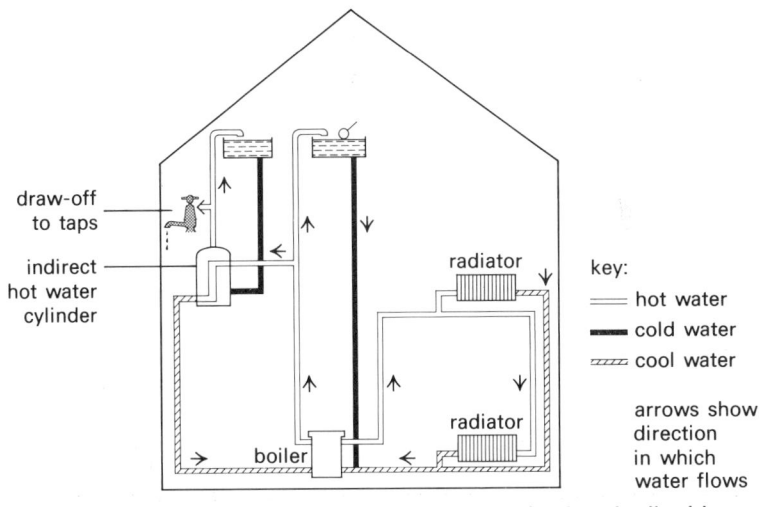

Gravity systems are slow to heat up so now an electrically driven pump is put in to speed up the flow of water.

## The indirect hot water cylinder

If a boiler is used in the system an indirect hot water cylinder has to be used because boiler water is too hot to come straight out of the tap, and it would be very dangerous. The hot water from the boiler comes in at the top, passes through the coil, and flows out at the bottom back to the boiler. The cylinder is fed with cold water, and after it has been heated it will pass out of the top of the cylinder to the taps in the house. Radiators use the hot water directly from the boiler.

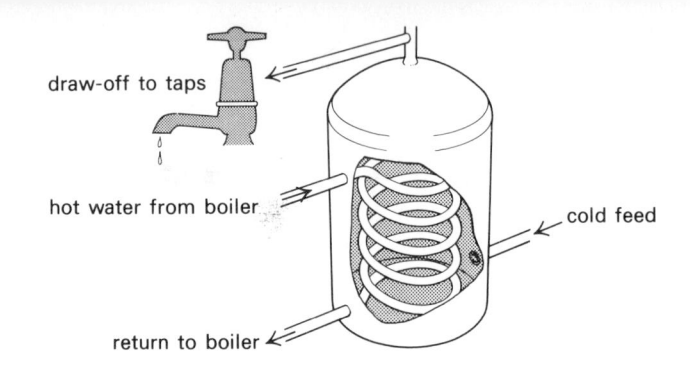

## Heating the home

Heaters mainly give out heat through convection, and radiation, and you will find that some are called convectors and some radiators. In fact they will do both. The hot water radiator gives off heat radiation into the room, whilst some hot air is convected upwards.

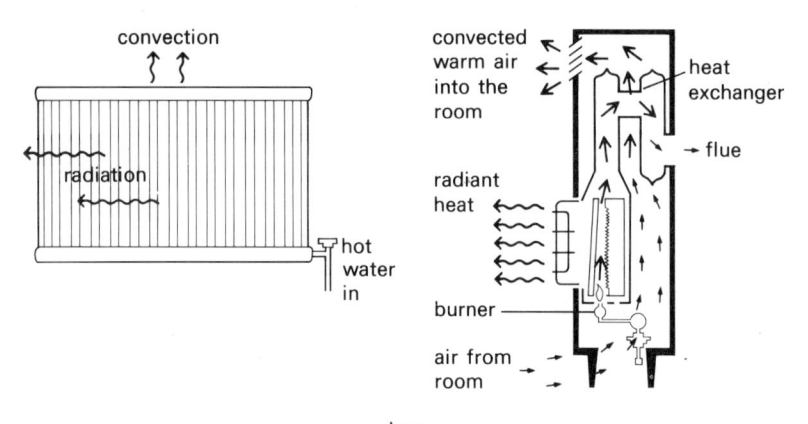

Hot water radiator

Radiant convector gas fire

key:
→ direction of movement of cool air
→ direction of movement of warm air

In the radiant convector gas fire the gas burner heats up the fireclay element which gets red hot and then radiates heat into the room. The cold air which is sucked up from the bottom is heated by the heat exchanger, and is then convected up into the room.

An electric fire has the heating element placed near the focus of the curved reflector so that the heat is radiated strongly out into the room. This is the same idea as the electric lamp and curved reflector in a torch or car headlamp.

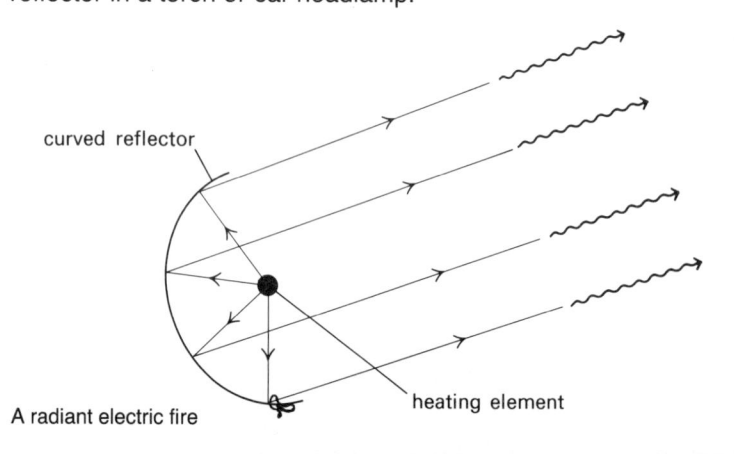

A radiant electric fire

### ACTIVITY D.1
1  Write a description of your own house heating system for rooms and hot water. Imagine that it is for your parents or brothers or sisters who do not know very much about heating, and use plenty of diagrams to help your explanation.
2  Say whether you think you have a good system or not, and give your reasons.

### ACTIVITY D.2
Make a survey of home heating. Use a table with the following types of heating systems: oil; coal; gas; solid fuel; electric central heating; electric fires; paraffin stoves. Include any others not on this list.